THE CITY LIBRARY
SPRINGFIELD, (MA) CITY LIBRARY

DISCARDED BY
THE CITY LIBRARY

Emily's
Place for
Children

JAN 27 2020

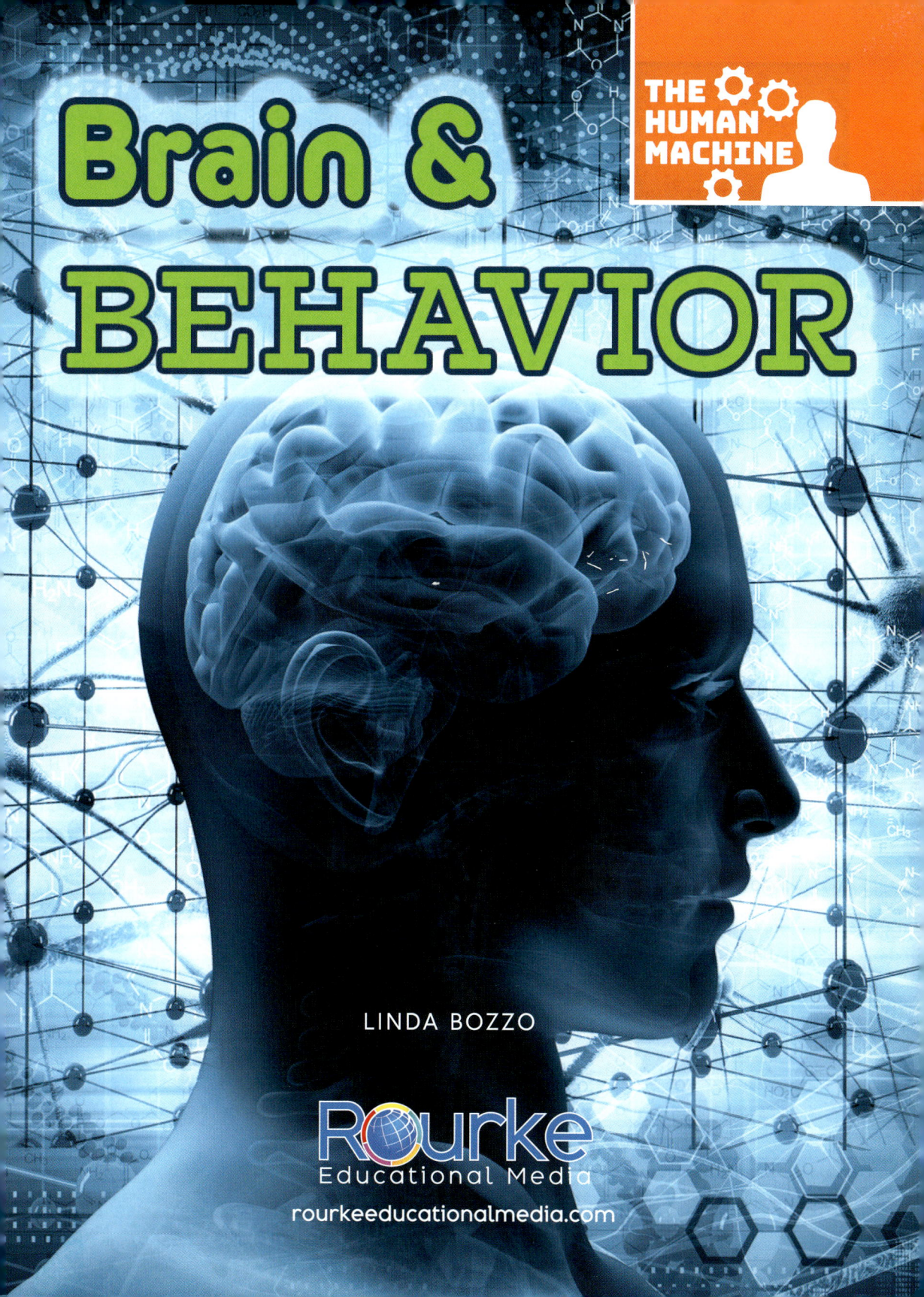

Before & After Reading Activities

Before Reading:

Building Academic Vocabulary and Background Knowledge

Before reading a book, it is important to tap into what your child or students already know about the topic. This will help them develop their vocabulary, increase their reading comprehension, and make connections across the curriculum.

1. Look at the cover of the book. What will this book be about?
2. What do you already know about the topic?
3. Let's study the Table of Contents. What will you learn about in the book's chapters?
4. What would you like to learn about this topic? Do you think you might learn about it from this book? Why or why not?
5. Use a reading journal to write about your knowledge of this topic. Record what you already know about the topic and what you hope to learn about the topic.
6. Read the book.
7. In your reading journal, record what you learned about the topic and your response to the book.
8. After reading the book complete the activities below.

Content Area Vocabulary
Read the list. What do these words mean?

circuits
digestion
grieving
imbalance
mass
membranes
organ
psychiatrist
react
spinal cord
x-ray

After Reading:

Comprehension and Extension Activity

After reading the book, work on the following questions with your child or students in order to check their level of reading comprehension and content mastery.

1. What does your brain control? (Summarize)
2. Why do people behave in different ways? (Infer)
3. How do neurons communicate with each another? (Asking Questions)
4. How does your brain tell your muscles to move? (Text to Self Connection)
5. How can brain research help people now and in the future? (Asking Questions)

Extension Activity

Choose a brain imaging method and research how it allows neuroscientists to see inside the living human brain. How does this method help neuroscientists? Create a table with three columns. Compare how scientists studied the brain 100 years ago, how we study the brain today, and imagine how we might study the human brain in the future.

TABLE OF CONTENTS

Brain Power	4
A Look Inside	12
Neuron Network	18
The Way We Behave	22
Imagining the Future	26
Glossary	30
Index	31
Show What You Know	31
Further Reading	31
About the Author	32

BRAIN POWER

Your brain is the most complex **organ** in your body. Weighing about 3 pounds (1.4 kilograms) in adult humans, it's one of the body's biggest organs. It controls the beating of the heart, **digestion** of food, and the workings of the lungs. But that's not all. Our brain helps us to remember. It controls our senses and gives us our unique personalities.

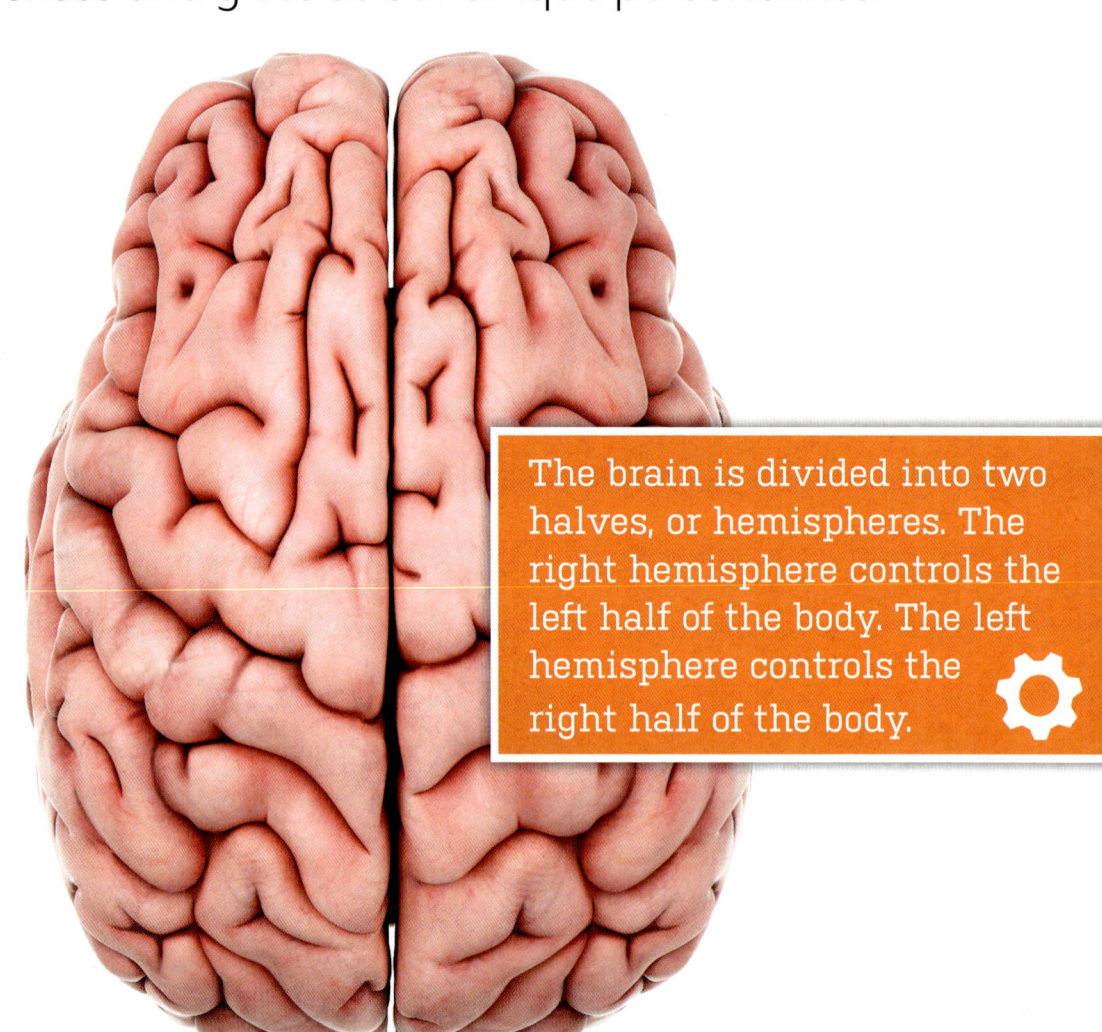

The brain is divided into two halves, or hemispheres. The right hemisphere controls the left half of the body. The left hemisphere controls the right half of the body.

MAIN PARTS OF THE BRAIN

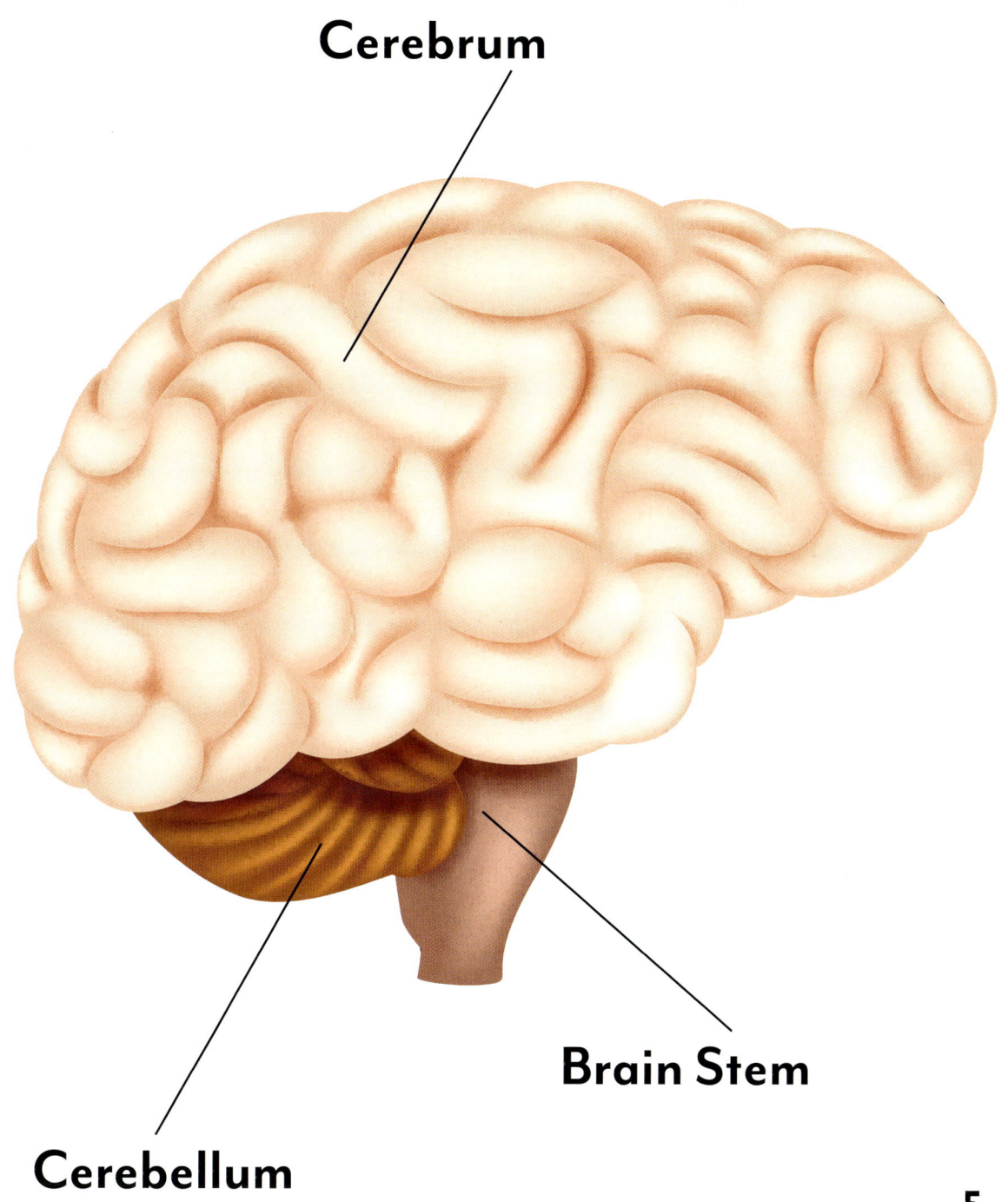

This **mass** of fat and protein that looks like a blob of gelatin is very delicate. It needs protection! The skull and three **membranes** called meninges protect the brain. The spaces between these membranes are filled with fluid. This fluid helps cushion the brain from damage.

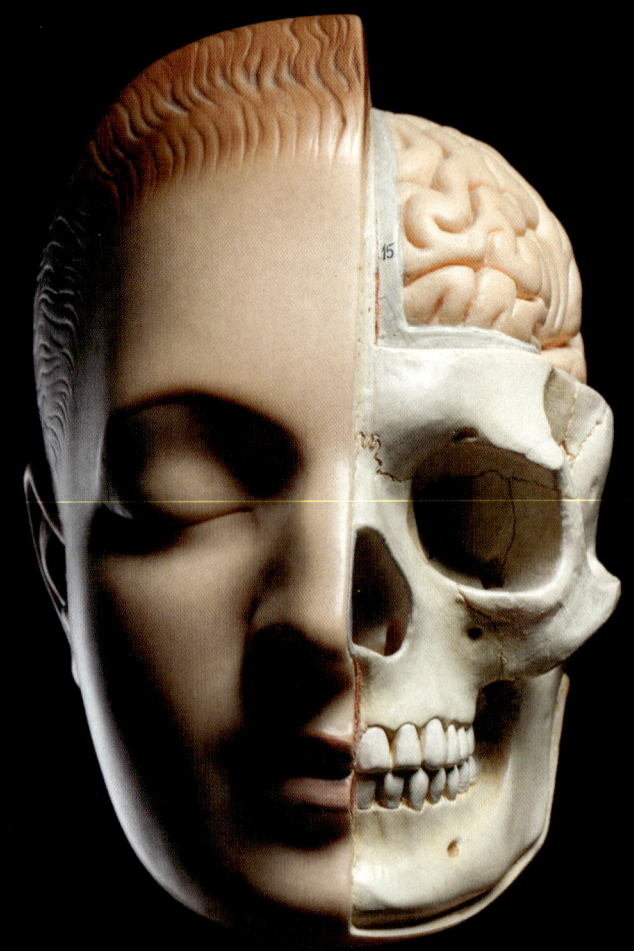

The brain is part of a larger system known as the nervous system. The nervous system includes the brain, the **spinal cord**, and the nerves that link them with all parts of the body. Using incoming and outgoing signals, the nervous system controls all of our body's activities. It also helps keep us safe.

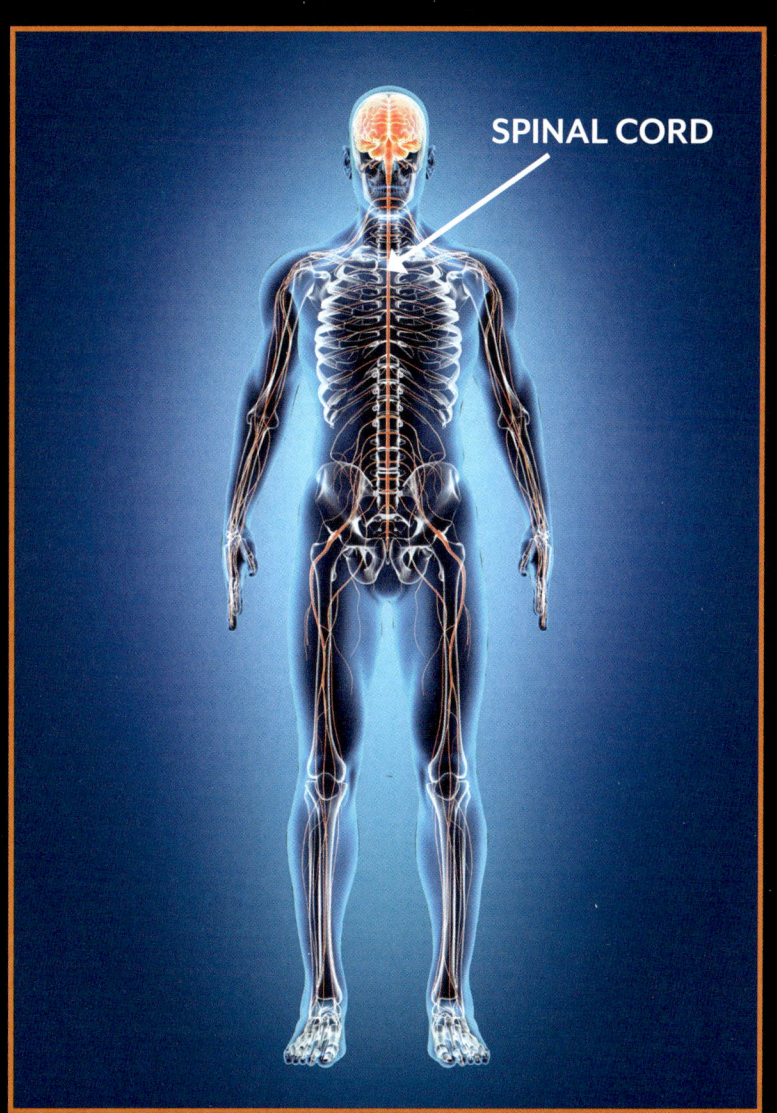

SPINAL CORD

For example, you see a car speeding towards you. A message is sent to your brain. This message travels down your spinal cord and to the nerves in your legs telling them to move quickly to avoid being hit.

The Cranial Nerves

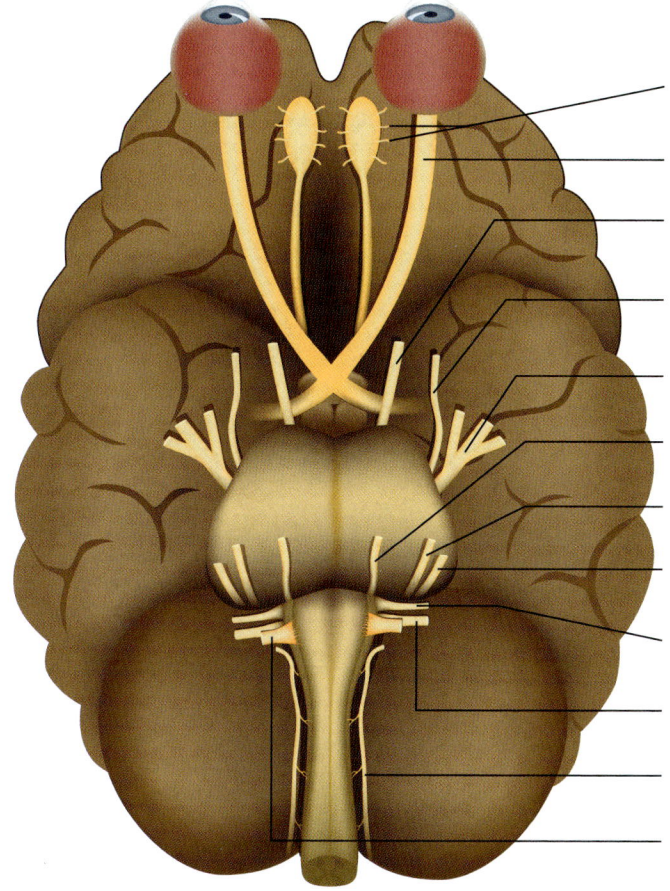

- Olfactory nerve fibers (I)
- Optic nerve (II)
- Oculomotor nerve (III)
- Trochlear nerve (IV)
- Trigeminal nerve (V)
- Abducens nerve (VI)
- Facial nerve (VII)
- Vestibulocochlear nerve (VIII)
- Glossopharyngeal nerve (IX)
- Vagus nerve (X)
- Accessory nerve (XI)
- Hypoglossal nerve (XII)

Cranial nerves come from the brain. They control muscles and organs above our shoulders, such as those in our eyes, tongues, ears, mouths, and lungs. Spinal nerves come from the spinal cord. They control the muscles in our arms and legs and organs such as our stomach and intestines.

The study of the nervous system is called neuroscience. Behavioral neuroscience is the study of the nervous system and how it relates to behavior.

Why study the brain and behavior?

Scientists hope to discover more about how and why people behave the way they do.

You might wonder how people can have the exact same experience but have different reactions. You can blame some of it on the brain!

One person shakes from fear of heights. Another screams at the sight of a snake slithering through the grass. Do you cry during a sad movie? Some people may. Others may not. These differences in behavior make us individuals.

A LOOK INSIDE

Studying the brain at work helps scientists understand how and why individuals' brains **react** differently to certain situations. Until the **x-ray** was discovered in the late 1800s, doctors and scientists weren't able to see inside a body without cutting it open. It's true! But we've come a long way. Now, modern tools help us study how an individual's brain might affect a person's health, memory, or behavior, among other things.

In 1895, German physicist Wilhelm Conrad Röntgen (1845–1923) made the accidental discovery of x-rays.

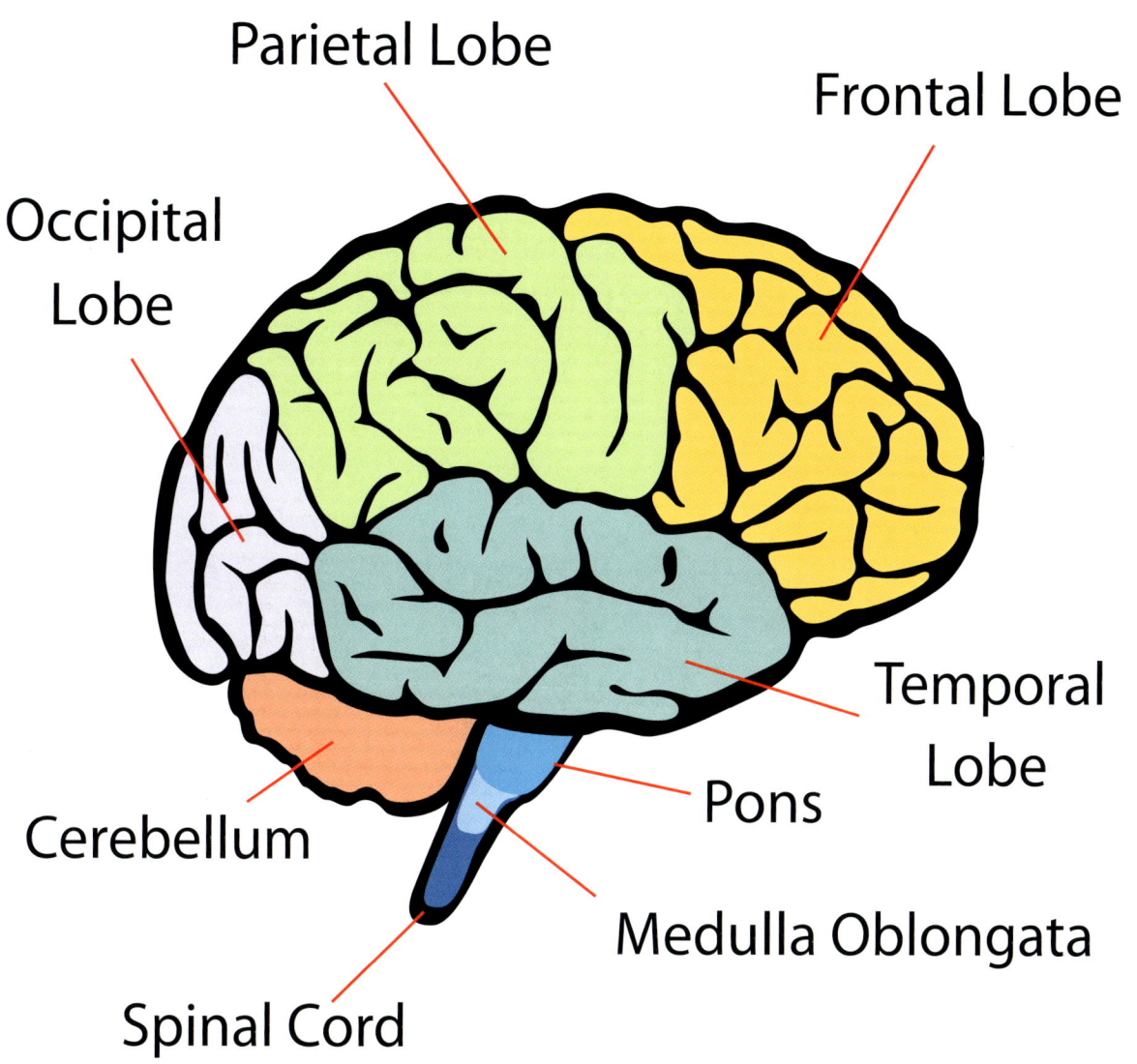

The spinal cord is a tube-like structure filled with a bundle of nerves and cerebrospinal fluid, which protects and nourishes it. The spinal cord is about an inch (2.54 centimeters) across at its widest point and about 18 inches (45.72 centimeters) long.

Modern imaging tools allow us to look inside a living brain. What can we learn from exploring the brain? Computed tomography (CT) imaging teaches us about the structure of the brain. It is more detailed than normal x-rays. Functional magnetic resonance imaging (fMRI) and positron emission tomography (PET) scans use blood flow to help us see how the brain works.

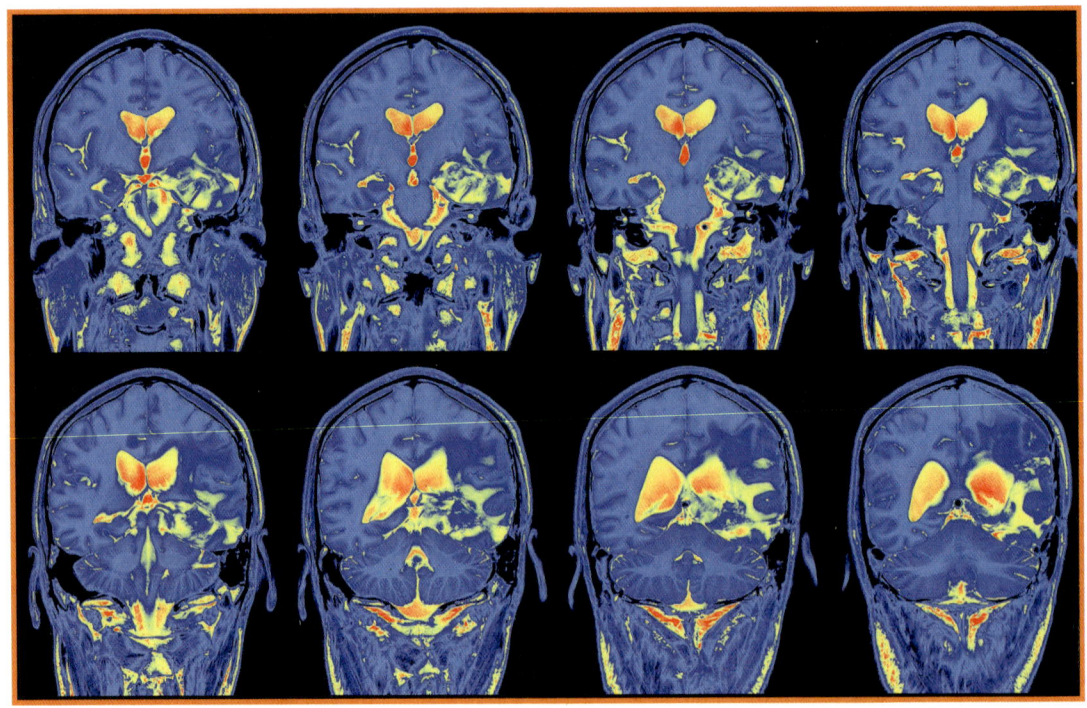

Doctors use fMRI scans to help understand which regions of the brain are linked to functions such as speaking or walking.

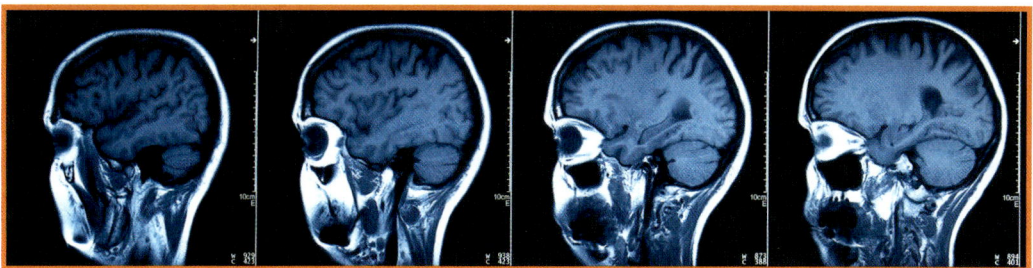

Using a computer and rotating x-ray machine, CT scans can produce numerous images of slices of the brain.

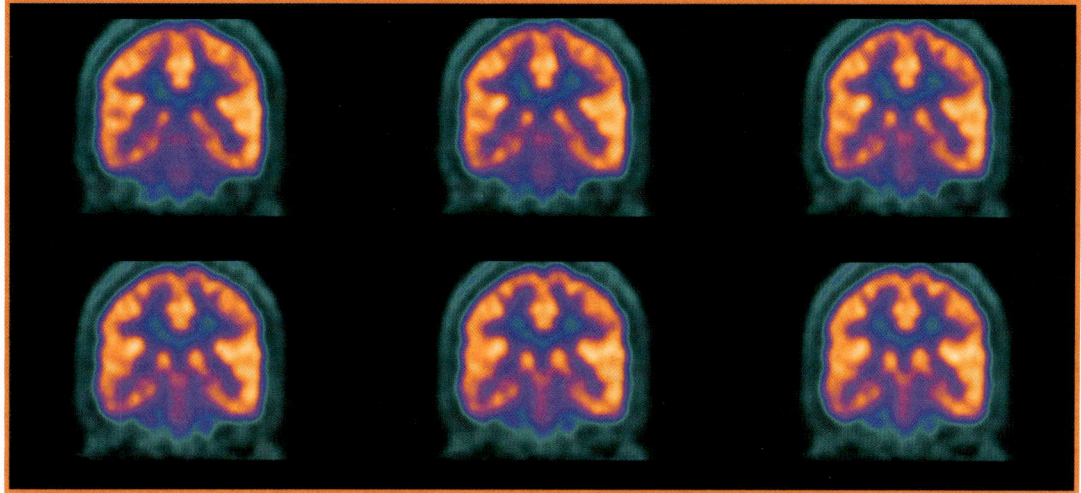

Small amounts of radioactive materials called radiotracers, a special camera, and a computer are used in PET scans to create images like these of the brain.

In 1848 Phineas Gage, the foreman of a road-building team, was injured when an explosion blew a metal rod up into his cheek and out the top of his skull. Amazingly, Gage survived for 12 years, but his character changed. The changes in his behavior supported an emerging medical theory that the frontal lobes of the brain are involved in human personality.

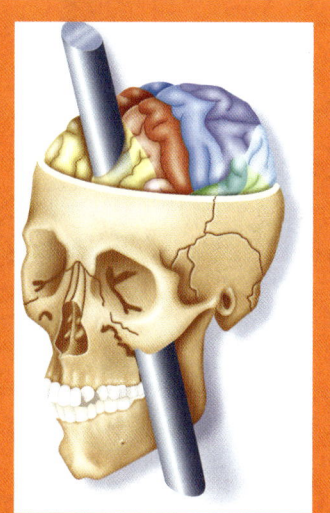

15

Brain-scanning methods can show which part of the brain is most active during certain tasks. Other tools show electrical activity in the brain. Still others use powerful magnets, radio waves, and a computer to make detailed pictures of the brain.

These tools and others play a role in helping us to learn more about the relationship between the brain and behavior.

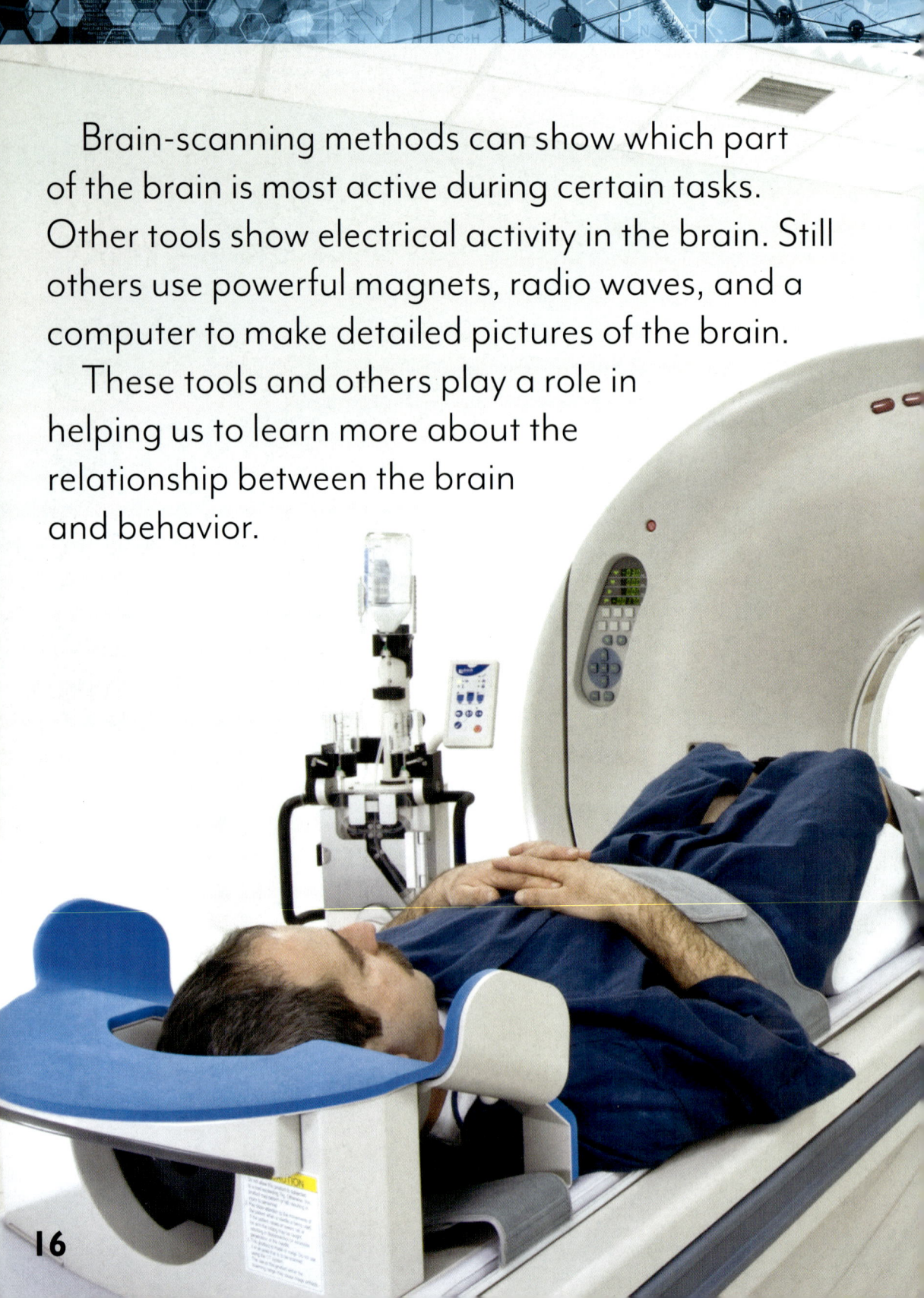

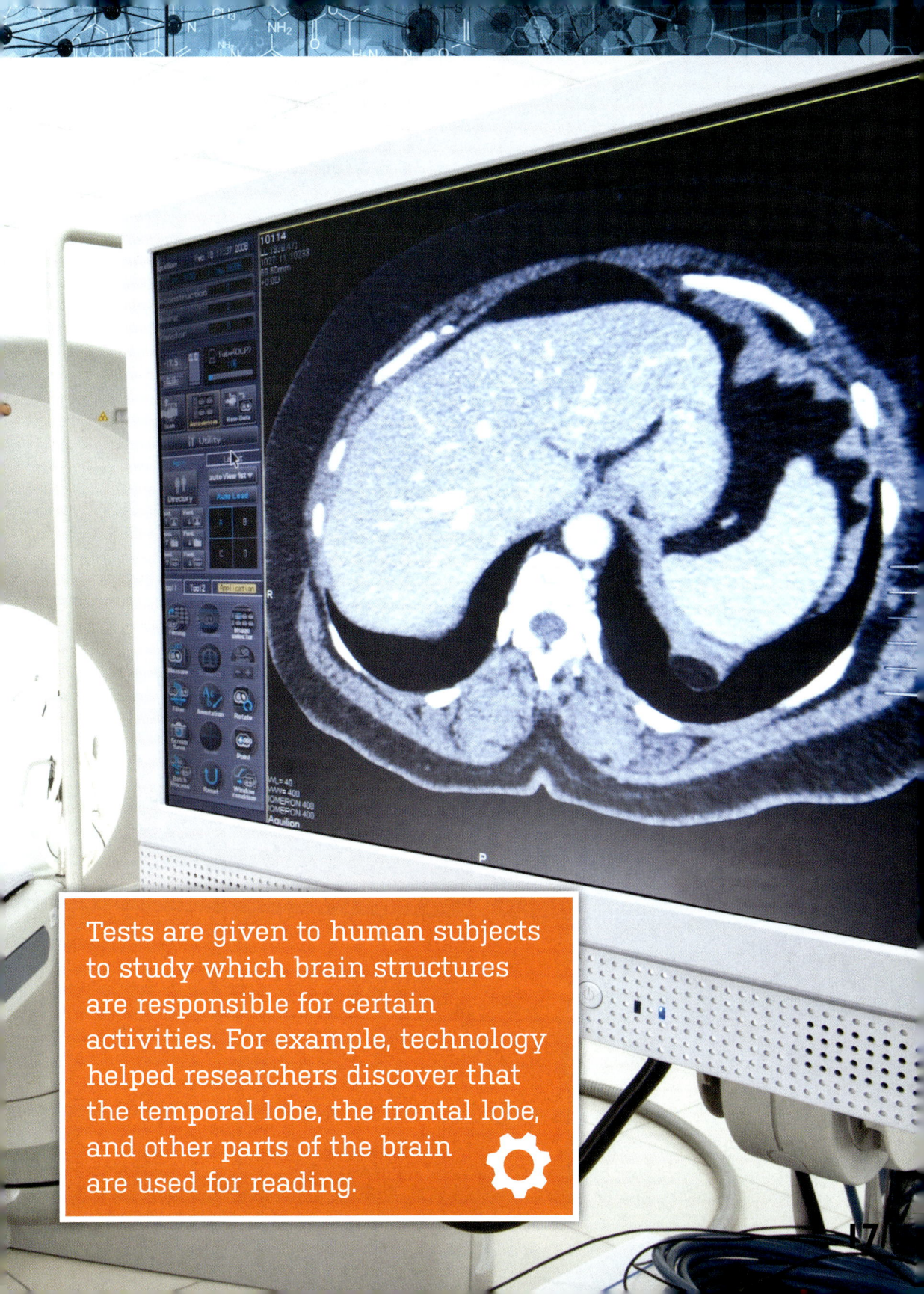

Tests are given to human subjects to study which brain structures are responsible for certain activities. For example, technology helped researchers discover that the temporal lobe, the frontal lobe, and other parts of the brain are used for reading.

NEURON NETWORK

To understand behavior, we must first understand the brain. We know that special nerve cells called neurons receive, carry, and pass on high-speed signals to other nerve cells. In fact, an estimated 100 billion neurons are found in the brain. Grouped together, these neurons communicate with one another by means of electricity and chemicals. Every second, trillions of signals are traveling throughout these **circuits** in your body. In part, it's these signals that relate to behavior.

Neurons, as well as some connections, are already present in the brain of a newborn baby. As the young brain develops, so do the connections. As part of normal brain development, the brain will get rid of connections that are rarely or never used.

Neurons are important cells. One of their jobs is to send chemical messages, called neurotransmitters, throughout the brain and between the brain and the nervous system. While some chemicals are released, others are received. It's these chemical signals, as well as electrical signals, that let different parts of the brain talk with each other. Everything we do relies on this communication. When this process doesn't work correctly, experts think it may create an **imbalance** in the brain. This imbalance may change how a person feels and behaves.

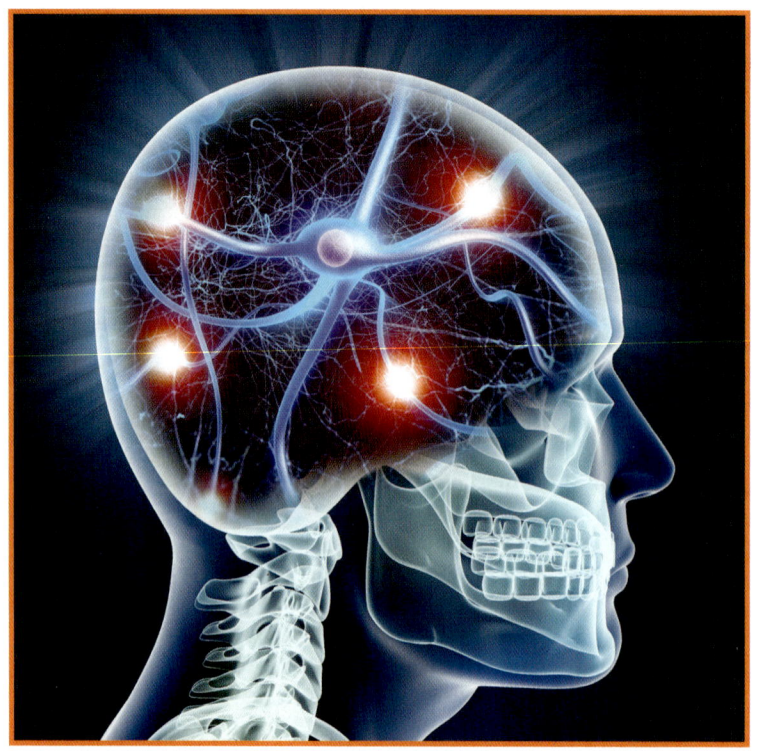

Your friends take your new cell phone. They start a game of catch with it. You politely ask for your phone back but instead it's accidentally dropped on the floor and breaks. What do you do? Shout? Cry? Stomp your feet? One person might start an argument. In the same situation, someone else might run away. Why the different behaviors? This is because each emotion sparks a slightly different pattern of activity in certain areas of the brain. Circuits in everyone's brains are wired differently. Therefore, we can't expect the behavior to be the same for any given emotion.

THE WAY WE BEHAVE

How do we determine what is normal behavior when everyone's personality is different? By looking at behaviors most people exhibit we can often consider what is not normal. For instance, **grieving** when you lose someone is considered normal behavior.

People with normal behaviors can manage their emotions and act appropriately in most situations. Abnormal behavior is when someone behaves in a way that may cause themselves or others discomfort or harm. Anger is normal. Destructive and violent behavior is not.

Psychologists are mental health professionals who study and evaluate behavior and mental processes. They use four criteria to identify abnormal behavior: violation of social standards, statistical rarity, personal suffering, and poorly adjusted behavior. Statistical rarity means something that is outside the range of average.

Though abnormal behaviors can't always be cured, they can be managed. This requires medical care. A common treatment involves a patient meeting with a **psychiatrist** in a private setting. Others might choose to meet in a group setting known as group therapy.

Often, more serious cases of abnormal behavior can be helped by treatment at an inpatient facility. This is where a patient moves into a facility and receives treatment for an extended length of time. Less serious cases might be treated at an outpatient facility where they are cared for during the day but don't spend the night.

Medications work well as a treatment for some people with conditions that result in abnormal behavior. These must only be taken while under the care of a doctor.

IMAGINING THE FUTURE

Technology that can read a brain's signals and send a simple signal that means "yes" or "no" is already being used. Researchers are working to make it possible to send a signal from a brain directly to a computer. By connecting a brain to a computer with a camera, someone who lost an eye may be able to see again. Imagine what other types of things we may be able to do with our brains in the future!

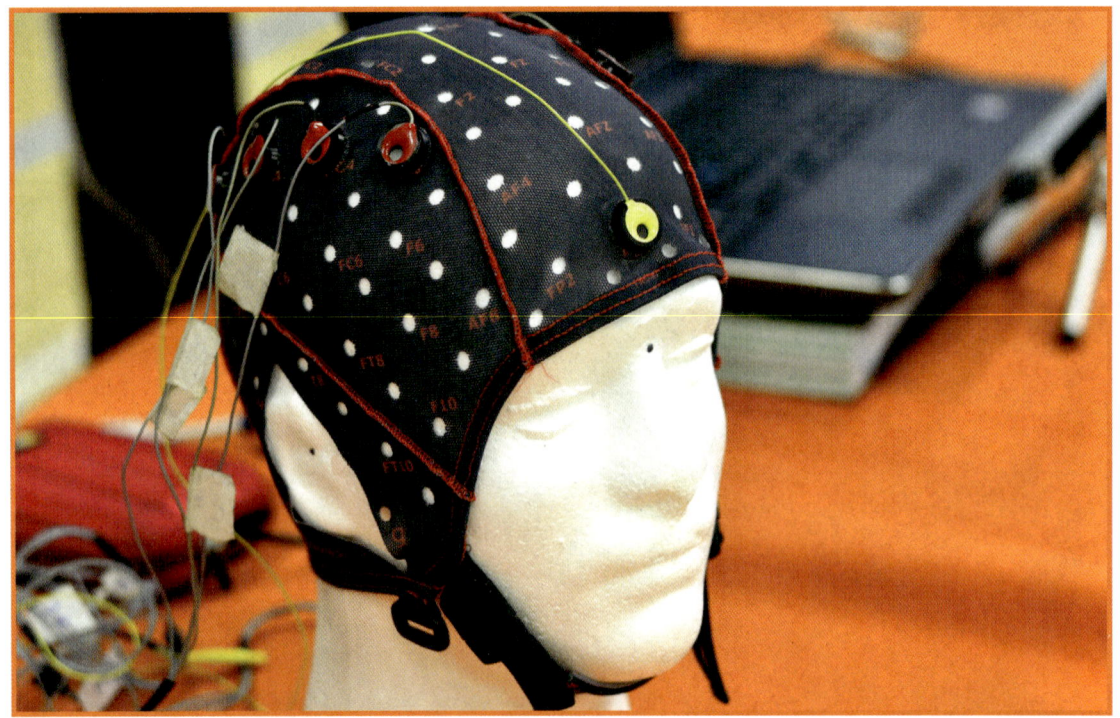

U.S. president George H.W. Bush proclaimed the 1990s "The Decade of the Brain" to bring attention to the importance and benefits of brain research. The Library of Congress and the National Institute of Mental Health of the National Institutes of Health worked together to advance the cause.

I, George Bush, President of the United States of America, do hereby proclaim the decade beginning January 1, 1990, as the Decade of the Brain. I call upon all public officials and the people of the United States to observe that decade with appropriate programs, ceremonies, and activities.

George H. W. Bush

So if you think we know everything about the brain, think again! While the brain is the most powerful organ in the human body, it's also the most mysterious. Even though scientists have studied the brain and behavior for decades, there is still much we don't know. However, through new inventions, tools, and technology scientists are learning more and more about the brain. They are also learning more about the vast number of jobs it performs.

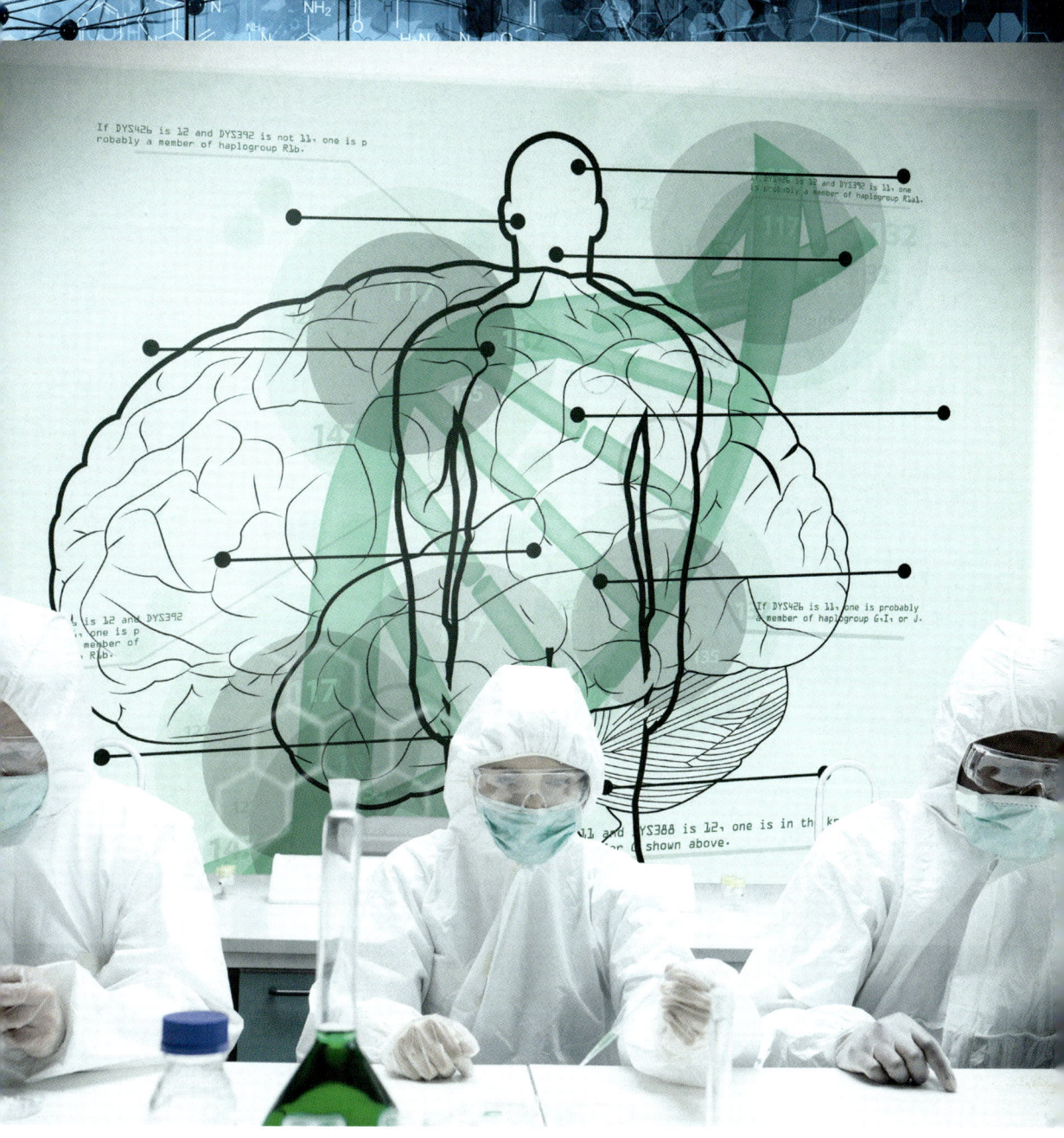

New discoveries are happening every day. Scientists try to guess what will come next. With gaining a better understanding of the brain and behavior, who knows what the future may hold!

GLOSSARY

circuits (SUR-kits): circular routes

digestion (duh-JESS-chuhn): the process of breaking down food in the stomach and other organs so that it can be absorbed into the blood

grieving (GREEV-ing): feeling very sad, usually because someone you love has died

imbalance (im-BAL-ens): a lack of balance

mass (MASS): a lump or pile of matter that has no particular shape

membranes (MEM-brayns): very thin layers of tissue or skin that line or cover certain organs or cells

organ (OR-guhn): a part of the body that does a particular job

psychiatrist (sye-KYE-uh-trist): a medical doctor who is trained to treat emotional and mental illness

react (ree-AKT): respond or behave in a particular way in response to something

spinal cord (SPY-nuhl KHORD): a thick cord of nerve tissue that starts at the brain and runs through the center of the spinal column carrying impulses to and from the brain and links the brain to the rest of the nerves in the body

x-ray (eks-ray): an invisible high-energy beam of light that can pass through solid objects used to take pictures of teeth, bones, and organs inside the body

INDEX

behavior 10, 11, 15, 16, 18, 21, 22, 23, 24, 25, 28, 29

cranial nerves 9

emotion(s) 21, 23

Gage, Phineas 15

imaging tools 14, 15

medications 25

nervous system 7, 10, 20

neurons 18, 19, 20

neuroscience 10

neurotransmitters 20

protection 6

senses 4

technology 17, 26, 28

treatment 24, 25

SHOW WHAT YOU KNOW

1. How is the brain protected?
2. What are some of the tools that help scientists study the human brain?
3. How do different parts of the brain communicate with each other?
4. What is the difference between normal and abnormal behavior?
5. What are some ways abnormal behavior can be managed?

FURTHER READING

Rose, Simon, *The Nervous System*, Weigl, 2015.

Roberts, Alice Dr, *The Complete Human Body,* 2nd Edition, DK Publishing, 2016.

National Geographic Kids, *Weird But True Human Body: 300 Outrageous, Facts About Your Awesome Anatomy*, National Geographic Children's Books, 2017.